Imagine this...

Stories Inspired by Agriculture 2009

J.D. Ortiz
Nicole Groteguth
Claudia Lopez
Anna Harris
Ciara Chiesa
Emma Morris
Braden Whitehouse

California Foundation for Agriculture in the Classroom

Walmart

Publisher's Cataloging-in-Publication Data

Ortiz, J.D.

Imagine this : stories inspired by agriculture 2009 / J.D. Ortiz, Nicole Groteguth, Claudia Lopez, Anna Harris, Ciara Chiesa, Emma Morris, Braden Whitehouse.

p. cm.

ISBN 978-0-615-34893-3

Summary: 2009 winning stories from the California Foundation for Agriculture in the Classroom, Imagine this... Story Writing Contest written by third through eighth grade students.

[1. California—Social life and customs—Fiction. 2. Farm life—California—Fiction. 3. Food—Fiction. 4. Short stories. American.] I. Groteguth, Nicole. II. Lopez, Claudia. III. Harris, Anna. IV. Chiesa, Ciara. V. Morris, Emma. VI. Whitehouse, Braden. VII. Title.

PZ5 .I326 2010
813.54—dd22 2010921500

California Foundation for
Agriculture in the Classroom
2300 River Plaza Drive
Sacramento, CA 95833
(916) 561-5625 • Fax (916) 561-5697
(800) 700-AITC (2482)
www.cfaitc.org

For California farmers and ranchers.
May you continue to produce the fresh, healthy
and affordable food that feeds us, fiber that
clothes us and materials that house us.

Table of Contents

Introduction

"Whether students live in the city or the country, they are surrounded by the world of agriculture. Turf is grown for lawns, fish are raised to stock lakes and genetics are used to improve the crops and animals that are our food. Not only is it important to teach about agriculture—it's essential. We eat it, live it, breathe it and are a part of it."

—Suzanne Squires, Santa Barbara County educator, grades 7-8

For 16 years, California Foundation for Agriculture in the Classroom has coordinated the *Imagine this...* Story Writing Contest to create agricultural awareness through reading, writing and the arts while meeting Content Standards for California Public Schools. By understanding how and where their food is grown, children and their parents will be more inclined to make healthier, more nutritious choices at the supermarket, contributing to stronger bodies and sharper minds.

The 2009 *Imagine this...* Story Writing Contest winners were selected from more than 10,000 stories about agriculture written by 3rd through 8th grade students throughout the state. The *Imagine this...* book showcases seven tales that are brought to life by the illustrations of Sacramento area high school art programs. We hope that you enjoy these stories, and that they will inspire you to share your appreciation for agriculture with the people in your life.

The Water Waster

By J.D. Ortiz
3rd grade, Kerman-Floyd Elementary School
Margaret Nichols and Michelle Karagozian,
Teachers – Fresno County
Illustrated by Inderkum High School

In the small town of Kerman, there lived an 8-year-old boy named Carter, but his family called him "the water waster." Carter was the youngest in his family so he thought that everyone always picked on him. He didn't really pay attention to the names that his brother and sister called him.

Carter did have a really big problem. His problem wasn't just once in a while—it was every single day, actually a few times during each day. He took long showers. Sometimes he would even take 40-minute showers. Most of the time he wouldn't even be in the water, he would just stand on the side of the shower and daydream as the water ran down, hitting the shower walls. His mom would yell at him, "Hurry up and get out, you water waster!" But Carter wouldn't listen, he couldn't hear what she was screaming when the water was running, but he could tell she wasn't happy.

At night he would brush his teeth and he would leave the sink faucet running as he took his time to clean every single tooth in his mouth. His sister would glare at him and yell, "Turn off that water, you water waster!"

One of his chores was to water the grass during certain times of the day but he would look for bugs, play basketball or just hang out in Lala Land while he was watering the cement driveway. His brother would yell at him, "Turn off that water hose, you water waster!"

Carter didn't understand why everyone was making such a big deal about leaving the water on. He didn't know what their problem was, and why they were trying to save water.

Finally his mom decided to sit him down and explain to him why saving water was so important. She told him that wasting water was not good for California because there is a **drought**. His mom said, "Because it doesn't rain a lot and because farmers need water to grow food so we can eat, we need to stop being water wasters. Plus California has a lot of wildfires, and we use tons of water when we have to fight these fires to protect people."

Carter really didn't understand how California was running out of water since water was everywhere, so he just thought she was making up stories. Carter decided his mom was trying to get him to take shorter showers so they wouldn't be late to school every morning.

That very same night Carter went to bed. But this night he did not sleep well. He had a terrifying dream; it was so bad that it felt real. In his dream, all the dried-up plants came to life. All the thirsty tomatoes, oranges, pistachios, apples and grapes were chasing him. They were dry and wrinkled, (you know how when you get out of the pool and your fingertips are all shriveled up), well in Carter's dream they all looked like that. The plants were very mad at him and they were yelling, "You water waster! You're stealing all of our water!"

Even the farmers were saying to the crops, "Get him!" They were chasing him with hay forks.

He woke up sweating and nervous.

That morning he went to his mom. "Mom, how long should I take a shower?" Surprised she said, "No longer than 10 minutes." Carter asked her for a timer and set it for exactly nine minutes.

That night, before he went to bed, Carter watched his sister brush her teeth and noticed that she didn't let the water run. Carter then brushed his teeth and did it exactly like his sister.

The next day when it was time to do his chores he asked his brother to help him water the grass. Carter thought that he could almost hear the grass drinking, gulping, and swallowing. They were probably so thirsty since that was really the first time he ever actually did it right.

From that day on Carter promised to be a water saver and not a water waster!

About the Author

J.D. Ortiz, age 8

The *Imagine this...* award-winning story, *The Water Waster,* was written by third grader J.D. Ortiz from Kerman-Floyd Elementary School in Fresno County. When first given the assignment to write about agriculture, J.D. was a little nervous because he didn't understand how agriculture related to his life; he thought it was only for farmers. After some research on water and the California drought, J.D. learned that agriculture is important to everyone. He thought about how he could help the world and decided to write a story based on someone he knew very well—himself! He used his experience as a water waster along with his imagination to come up with his story. He hopes his story about going from a water waster to a water saver will help others realize it is easy to conserve water!

About the Illustrator

Julia Wolfert
Inderkum High School – Natomas Unified School District
Art Instructor: Michelle Townsend

High school senior Julia Wolfert was excited to use the skills that she learned from her digital photography and graphic design classes to illustrate *The Water Waster*. Julia then decided to use this illustration assignment as her senior project, a requirement to graduate. She started the project by developing a storyboard to help organize her thoughts; she drew the illustrations on paper then loaded them onto a computer where she used digital shapes and colors to finish the illustrations. Initially she didn't understand very much about how water and agriculture were related, but through this project she learned a lot and believes that it has really opened her eyes to a new aspect of art and illustration, and has prepared her for a future in graphic design.

THE COMPOST WORMS

By Nicole Groteguth
4th grade, Gratton Elementary School
Pennie Segna, Teacher – Stanislaus County
Illustrated by Delta High School

"Stella, school today!" Momma worm said.

"I'm not ready today, Mom," Stella worm said. "Please give me, umm, how about ten hours."

"No," said Momma worm, "because this is your first day of school."

Stella had a great day at school. When she came home from school she asked her mom what they were having for dinner.

"We're going to have lettuce, banana peels, and pieces of the Earth Worm Times," said Momma worm.

"What? I thought we finished that last night!" said Stella worm.

"No honey, remember your brother left some in the dirt pile," said Momma worm.

The next morning, Momma worm called out, "It's breakfast time!"

"OK, I'm coming," Stella worm answered back. "What's for breakfast?"

"We are going to have lettuce, rotting apples, and don't forget banana peels, because that is just the right kinds of food worms love to eat." Momma worm told Stella that the food goes in the bloodstream and it is very good for worms.

Ten minutes later. "Bye Mom, I'll see you soon," said Stella worm.

At school the teacher told the class, "Today class, we are going to do fractions. Come gather around my desk and I will take a bite out of this half-eaten apple. Who can tell me how much is left?" The students all laughed as they watched. School was fun.

The teacher taught the worms all about their purpose on Earth. Worms help **cultivate** the soil by fertilizing and keeping it healthy for plants to survive. She mentioned that worms have been helping the Earth and plants for thousands of years.

Soon the teacher yelled, “Lunch time!”

“Uh oh,” cried Stella, “I already finished my lunch during snack time!” Stella began eating her math book because she was still very hungry. The teacher walked by and noticed her. The teacher made her write, “I will not eat my math book.”

Later at school, Stella found a friend named Penny worm. They had a sleepover that night at Penny's house. They dug day and night in the garden for fun. The worms also learned at school that this allows the plants to get air from the holes.

The two worms decided to have a picnic in the garden. They cleaned up the area by eating dead, old leaves. It soon became too sunny there so they decided to go to a shady park. They found that there were too many people at the park. They became afraid they would get stepped on. Then it began to rain. They quickly hid under a leaf because if they went down into a hole they could drown.

All these explorations made Stella worm and Penny worm very thirsty. So they drank some mud juice. Water is important because it helps worms make mucus that allows them not to dry out.

Penny remembered something she learned in school. "Stella, did you know that we have a very tiny brain?" stated Penny.

Stella cried "Nooo!"

Penny reassured Stella that they had enough brain power to control their bodies.

Then Penny said, "I'm hungry." So they ate some dirt rolls. The worms learned at school that everything they eat becomes digested and turns into rich fertile soil good for plants. They both wanted something sweet to eat. They made mud pie and ate it all. When worms eat dirt they also eat the tiny stones in the soil. The worms swallow these tiny stones along with food. The food then passes into the gizzard, which is made of strong muscles. The gizzard, with the help of the tiny stones, grind the food up making it useful to the intestines.

They ate until there was nothing left. Before they went to sleep they dug a while which made them safe and very sleepy.

On their way to school the next day, they bumped into a big root because they were talking and not looking at where they were going. They weren't looking because worms don't really have eyes. So, of course they couldn't see. After school they decided to play a game called "Find your way out." The game is played by digging out of buried holes in the dirt.

Stella and Penny grew to be great worm friends. They were very proud of the way they could help the earth by cultivating the soil so plants could grow and they would be contributing to everyone's needs. They dug and they played happily ever after.

About the Author

Nicole Groteguth, age 9

Nicole Groteguth, a fourth grader at Gratton Elementary School in Stanislaus County, is very proud of her award-winning story, *The Compost Worms*. Nicole confessed to being a bit scared about submitting her story for the *Imagine this...* Story Writing Contest, especially because she had not completed many stories from start to finish, but challenged herself to write about these earth-friendly invertebrates after studying worms in class. Nicole is very familiar with the world of agriculture and the many benefits that it provides, as her family grows hay and raises livestock in their rural community of Denair, located south of Modesto in California's Central Valley. She applauds her teacher and her classmates for writing about food, animals, and the wide variety of other agricultural themes for the contest, as she believes agriculture is a very important part of life that everyone should recognize.

About the Illustrators

Lauren Miranda, Rebekah Collins, Katrina Dominguez, Austin Daniels, Daniel Harding, Camille Burt, Ivete Rodriguez and Sarah Bills (not pictured)
Delta High School – River Delta Unified School District
Art Instructor: Denise Sanchez

Delta High School's digital media class worked as a team to illustrate the award-winning story, *The Compost Worms*. Although the school and much of the community is surrounded by agriculture, the students felt that after doing some research to prepare for illustrating the story, they learned even more about what is going on around them, especially about what happens deep within the soil. The students were surprised to learn how important worms are to the earth and agriculture. To begin this project, the group broke the story up by page number and each student was assigned a page to design. The students were constantly communicating with each other to make sure that all the pages flowed and the worms in the story looked the same on each page. The students sketched all of the illustrations with as much detail as they could. Finally, the digital media class used computers to enhance the details and colors of the finished illustrations for a fantastic final product.

Julio's Fig Adventure

By Claudia Lopez
5th grade, Fred Ekstrand Elementary School
Catherine Rojas, Teacher – Los Angeles County
Illustrated by Woodland High School

Jose, his wife Margarita and their son Julio own a farm in Madera, in the Central Valley of California. They moved there from a big city because they wanted to grow crops that would feed their family and other people.

On the farm they grow a lot of vegetables and fruit trees. Of all the things they grow and produce on the farm, Julio was interested in the figs the most. His grandmother used to talk about the farm she grew up on and all the fig trees they had.

He started to research figs and he found that the Spaniards brought figs to America in the 1500s and the missionary Fathers introduced them to California in 1759.

At the farm they have two different types of fig trees. One is amber color and called Calimyrna, the other has a dark purplish skin color and is called Mission—because that was the type planted by the Spanish Fathers.

Madera has dry, warm weather which is the best to grow and produce figs. The main crop ripens in late summer and produces the best figs.

Julio also learned that unlike other fruit trees, fig trees have no blossoms on their branches. The fig fruit is the flower of the tree, because flowers and seeds grow together in one mass.

Every day Julio helped his father do chores around the farm. During the summer, when school was out, Julio helped pick the ripe figs to sell at a stand on the side of the road, next to the farm.

One year, Julio heard that the county fair was coming to town. There was going to be a contest for the best **marmalade**. He came home so excited and asked his mom, "Could you help me make fig marmalade?"

His mom said, "Sure, why not?" Julio said, "We only have two more weeks to try the recipes."

The contest became a family project. Julio picked the figs early in the morning with Dad's help. He had two separate baskets, one for each kind; he knew that the different figs would produce different colored marmalades. Then, with Mom's help, they would start the process of making marmalade by washing the figs and cutting off the tips. They were supposed to put the figs in two big bowls, but his mom only had one. So he went into town and started to look in different stores until he found one. He was not sure about the size, but he bought it anyway. His mom thought it was perfect!

Once they cut the figs into fourths it was time to start cooking. Mom suggested making the two kinds of marmalade a little differently. The Mission figs would have water and sugar and the Calimyrna would have water, sugar and lemon juice. Julio and his mom put them in big pots over the stove. It was Julio's job to watch the pots and stir them every once in a while until they started to boil. Once they started to boil he had to reduce the heat and stir more frequently.

After about an hour and a half it was done. They poured the mixtures out of the pots and waited until they had cooled off so he and his mom could taste the marmalades. Julio thought the Mission had a rich dark color and the Calimyrna was a pretty amber color that he could almost see through. The taste? Well, he thought it needed to be a little sweeter. So they made another batch that came out just perfect. They poured the marmalades into a few special jars that were going to be the samples for the judges and the rest was canned in small jars to be sold.

He decorated labels for the jars, designed the booth with one area to display the produce from the farm to be sold and the other side for the marmalade.

Today was the day! The whole family woke up early, loaded the truck and headed to the fair. Julio set up his area of the booth with the jars for the judges on one side of the counter and the other side for the customers, with small containers so everyone could try it.

The judge started tasting; Julio was number seven out of fifteen. Soon after the tasting was finished, winners were announced. Julio won second place! He went up to get his trophy and ribbon.

When Julio came back he was a little disappointed but his mom and dad said, "You entered a contest for the first time and you won second place, we are very proud of you!"

About the Author

Claudia Lopez, age 10

Fifth grader Claudia Lopez from San Dimas, Los Angeles County, enjoys listening to her grandmother tell stories about what it was like growing up on a farm. It was one of those conversations that got Claudia thinking about agriculture and gave her the idea to write the award-winning story, *Julio's Fig Adventure*. Claudia is not new to the *Imagine this...* contest as she was recognized as a regional winner in 2008. She was excited to get the assignment to write another story for the contest because it gave her the opportunity to learn more about California agriculture and all the products we produce! Claudia has enjoyed writing stories so much that she wants to continue to write adventure stories for younger students.

About the Illustrators

Jordan Miller, Nick Christen, Zinthia Segoviano,
Jessica Landeros and Nicole Jackson
Woodland High School – Woodland Unified School District
Art Instructor: Dawn Abbott

Woodland High School's advanced placement art students each used their individual strengths to bring Julio and his story to life while illustrating *Julio's Fig Adventure*. Before the group could start illustrating they had to research geography, trees and figs to make sure the illustrations were accurate and true-to-life. The group met several times to brainstorm ideas before they began by hand drawing the graphics in pencil and using gouache, a type of paint, to finish the scene. The group learned a lot about working as a team to accomplish a common goal and believe they were successful because they worked together, listened to each others' ideas and let every person show his or her strength. They were honored to illustrate a story that was written by a younger student.

The Magic Green Tractor Visits the Dairy Farm

By Anna Harris
6th grade, Grenada Elementary School
Debbi Hoy, Teacher – Siskiyou County
Illustrated by Sheldon High School

Ms. Jersey is always taking us on crazy field trips. The latest field trip happened just last Friday. We were studying cows and all of the amazing products people get from them, when suddenly, Ms. Jersey shouted, "To the tractor, kids!"

"Not another field trip!" cried Donnie.

"Oh yes! The best way to learn about a dairy farm is to visit one!" replied Ms. Jersey.

So into the tractor we went. Zip, Zip, Zap, and all of a sudden we were at our local family-owned dairy farm.

The Duggar family has been operating this farm since 1854. Ms. Jersey asked Mr. Duggar to show us around the milking machines. We learned that it takes an average cow three to five minutes to give her milk. The extracted milk then passes through a strainer into a tank. It can be stored safely at around 42 degrees for approximately three days. He showed us how each cow had its own bed and access to food and water 24 hours a day. He told us that comfortable cows give more milk.

The milk truck was just arriving to transport the milk to a dairy factory for processing and **pasteurizing**. This is when the trouble began.

Ms. Jersey decided that we should hop back into the magic tractor and zap ourselves into microscopic size in order to hitch a ride in the milk tank with the milk to the factory. "We never zapped ourselves microscopic at our old school!" shrieked Alice.

But, we slid through the cooler vent anyway into the milk tank where we saw big chunks of **protein** and fat floating around. This would have been perfectly safe if only Donnie hadn't been **lactose intolerant**!

We helped him aboard a chunk of fat where he clung until finally we arrived at the factory. Ms. Jersey shouted, "Follow me, kids!" and we all slid down into a big **vat**. Poor Donnie was really suffering; he was holding his stomach, wailing for the bathroom.

Alice said, "Hold on, Donnie! It will be okay!"

The factory workers began separating the **curds** and **whey**. Thankfully we had hopped out and were standing on the rim of the tank by then.

The cheese makers added **rennet** to get the milk to **coagulate**. Then, starter **bacteria** converted the milk sugars into **lactic acid**. The rennet helped set the cheese into a strong texture. The soft cheese was drained, salted, and packaged for stores, right there in the factory. Ms. Jersey had to ask about making hard cheese because Parmesan was her favorite.

Donnie was really starting to feel better so the processors explained how hard cheeses can be heated to temperatures at about 100-130 degrees. Salt is added to the cheese for flavor and preservation. The cheeses are pressed into a mold. The harder the cheese, the more pressure it takes to shape it and squeeze out the moisture.

We discovered that some cheeses are ready right away and some are allowed to age, like Brie and Camembert. Donnie asked about what kinds of cheese he could eat with his lactose intolerance. They told him that aged cheese contains almost no **lactose** and cheddar cheese only has 5 percent.

"Amazing discovery, Donnie!" Ms. Jersey exclaimed.

"Cheese can also help people fall asleep faster and relieve stress because it has **tryptophan** in it," said Alice. "I never learned that at my old school!"

"Well kids, we better board the magic tractor and get back to our regular size before we crumble like feta!"

After a very educational day at the dairy farm and factory we were all glad to be back at school.

"What a gouda field trip," Ms. Jersey said.

"Ms. Jersey!" moaned the kids. And that was the end of our journey in our green tractor!

About the Author

Anna Harris, age 11

Anna Harris, a sixth grader from the Northern California town of Yreka, has always enjoyed learning about science fundamentals by reading Scholastic's book series *The Magic School Bus*, and decided to model her award-winning story after these adventure books. Despite growing up in rural Siskiyou County, Anna admits that her prior knowledge about agriculture was fairly limited. She says she learned a great deal while doing the research necessary to write this detailed and very scientific story about dairy processing. This talented eleven-year-old was especially surprised, and pleased, to learn that there are a variety of cheeses available for people, like her own mother, who are lactose intolerant. Anna believes it is important to integrate humor, conflict and interesting facts into her writing order to draw in readers, and was certainly successful in doing so in this clever and educational story.

About the Illustrators

Ba Xiong and Miyokang Kumangai
Sheldon High School – Elk Grove Unified School District
Art Instructor: Debbie George

Seniors Ba Xiong and Miyokang Kumangai have been friends since they were freshmen at Sheldon High School, and were excited to illustrate *The Magic Green Tractor Visits the Dairy Farm* as a final project together. Both students are using this illustrating experience as the basis for their senior projects. At first the artists thought it would be easy to illustrate a children's story, but quickly learned that it would be a more involved endeavor. The artists did a lot of research to make sure the illustrations were accurate. In doing so, they were surprised to learn that cheese can make you sleepy! Ba and Miyokang enjoyed the freedom they had to bring to life the characters and illustrations in the story. They drew visual inspiration from books, magazines, pictures, and previous projects they have worked on. They relied on each other's strengths throughout the project, learned a lot about agriculture and hope that their illustrations help others understand more about the process of making cheese!

Amy's School Garden

By Ciara Chiesa
7th grade, Gratton Elementary School
Rexann Jensen, Teacher – Stanislaus County
Illustrated by Delta High School

Once there was a seventh grade girl named Amy. She decided to start a school garden, but not just any garden; she wanted it to be beautiful and useful.

She said, "I will grow flowers, herbs, lettuce, carrots, cucumbers, broccoli, corn, onions, garlic, peas, and potatoes. I will be able to give pretty flowers to my mom, my teachers, and the principal. Also, the school will have healthy and nutritious food to eat that is fresh from the garden."

Her class had just finished learning about the importance of healthy foods in a person's diet. They had also learned about the different food groups and that a person should eat a lot of vegetables each week.

Amy's idea for a school garden soon caught on and more and more students wanted to help Amy with her project. One of her friends suggested that they should try **composting** to add **nutrients** to the soil. They looked up ideas on how to compost from the library.

Some of her classmates picked up leaves at recess and some of her friends collected food scraps during lunch from the students to make compost. After the compost was ready, they added it to the soil. They mixed it in so the soil was soft and ready to plant. Then Amy and her classmates dug holes and planted the seeds.

Almost every day there was something to do in the garden. Amy and her classmates pulled weeds, watered the seeds, and made sure they received plenty of sunlight. Once a month Amy even gave them some **fertilizer**.

One day while she was watering, Amy's friends Lizzy, Mila and Katie came by to see her and her garden. Amy told her friends that she thought her garden was taking too long to grow. She wondered if she had done something wrong.

Lizzy said to give the plants more water. Mila suggested that Amy should give them more fertilizer. Katie told Amy she read that plants grow better when a person sings to them. Amy decided to ask her teacher for some advice too. Her teacher told her to just be patient. Amy chose to go with her teacher's suggestion.

So over the next three weeks, Amy watched and waited for the plants to grow. One day a little green sprout sprang out of the ground. The next day a few more green **seedlings** popped through the soil. Amy was so excited about her garden growing that she called all her friends and told them about her new discovery. The whole school was excited about the garden now. Lots of students took turns tending to the garden duties, and the plants continued to grow.

Vegetables started appearing. After a few more weeks, they started maturing until finally the day came when the vegetables were ready to be picked. Soon the garden had enough produce that Amy's class was able to make a really good soup and salad for the whole school to try. Some kids had never tried vegetables like these, but because they had helped in the garden, they wanted to see how they would taste. Everybody seemed to like the soup and salad. Some parents called to get the recipe because their children said it was delicious and nutritious.

From that day, Amy realized that while it is smart to have good soil, plenty of water, and fertilizer, one of the most important things of a garden is patience, patience and even more patience. Some things are worth waiting for and Amy and her classmates agreed a garden is worth the work and the wait.

About the Author

Ciara Chiesa, age 13

Ciara Chiesa is no stranger to the world of agriculture; her family has farmed walnuts, almonds and peaches for five generations in the Central Valley community of Hughson, California. Ciara crafted *Amy's School Garden* from her own experiences helping her grandparents in their large garden, and after participating in the revitalization of her school's garden at Gratton Elementary. As the garden grew, Ciara noticed that many of her classmates who were normally reluctant to try many types of fruits and vegetables were suddenly eating them fresh from the garden! This inspired Ciara's well-crafted storyline, and allowed the 13-year-old author to have fun with the writing process. Ciara is very excited to see her story illustrated and printed, and is pleased to be able to share the story of agriculture's important role in feeding the world through this creative story about school gardens.

About the Illustrators

Back row: Drake Hart, Eduardo M. Corona Jr., Jonathan Williams, Jeffrey Kreidler and Joseph Ore
Front row: Jaira Garcia, Adriana Mandujano-Rodriguez, Josie Rodriguez and Michelle Arceo
Delta High School – River Delta Unified School District
Art Instructor: Denise Sanchez

The student artists that illustrated *Amy's School Garden* are from Delta High School in the Sacramento River Delta. The talented group has many specialized skills they showcase in the illustrations for the story. With nine students in the group, and many ideas on how to illustrate, they learned a few lessons in compromising, working together and working for someone else. The students wanted to make sure that there was continuity in the illustrations, so they broke the group up into smaller teams to take on illustrating particular elements of each scene: one team worked on drawing the settings, another worked on vegetables, and one team worked on illustrating Amy and her friends. This group also used computer programs to add final touches to the illustrations. The final product is one that the students are very proud of!

By Emma Morris
8th grade, Scott Valley Junior High School
Tracy Dickinson, Teacher – Siskiyou County
Illustrated by Woodland High School

I shoved my foot into my boot and tucked in my jeans. Grabbing my coat from the hook, I slipped through the door behind my dad. I gasped as I drank in the freezing morning air. It must have snowed in the few hours I was asleep because the ground was blanketed with a thin, white sheet. I stood there for a second, awed by the beauty of it all, and then slipped my coat on and ran to catch up with dad.

“Hey,” said Dad, “It’s beautiful huh? But this is terrible weather to lamb in, so we better go check on the **ewes**.”

“Okay, I hope there are more babies!” I said excitedly.

We walked through the gate and into the frosty field. Most of the sheep were sleeping, but one was standing by itself. “Dad,” I said, “over there.”

“Yeah, let’s go check it out.” We walked over to the ewe and found she had three babies on the ground next to her. They were shivering violently so I bent down and put my finger in one’s mouth to check its temperature... cold, not a good sign.

“Em,” said Dad softly, “Grab a lamb and help me coax this ewe into the barn.”

“Okay,” I replied, picking up one of the babies gently. It took us about 10 minutes to get the ewe into the barn, and once we were in I set the lamb down on the soft, clean straw. The baby grunted quietly, still shivering.

“Emma,” said Dad, “Plug in the hairdryer and start trying to warm these little guys up.” I nodded and grabbed the battered hairdryer we use to warm up the lambs. I picked up the babies and set them together on a warm towel, and proceeded to blow hot air on their shaking bodies. After about 10 minutes, they started to warm up and tried to stand on their wobbly legs. “They're warm Dad,” I said.

“Okay,” he said, “Let's get them some starter fluid.” “Starter Fluid” was Dad's name for some protein supplement we give lambs to get their blood flowing. I grabbed the bottle from the shelf, and squirted the thick substance into their mouths. They gulped it down, and continued trying to stand up.

“You’re going to have to pick one to take inside,” said Dad, “This ewe can’t take care of three babies herself.” I studied the babies. Two of them had stood up and were trying to get milk from their mom. The third was still struggling.

“I guess it’s going to be that one,” I sighed.

“I agree,” said Dad, “Why don’t you take him in, and warm up some milk. He’s going to need something in its stomach before too long. I’ll handle the rest of this.”

"Okay," I said picking up the baby and wrapping it in a towel. I carried the lamb across the yard, and then into the house. The warmth hit me and made my cheeks burn. I set the baby down in a box we had ready-made for occasions like this one. The box had fresh towels and a hot water bottle. Then I walked over to the sink to get some milk ready. First I grabbed the powder from the cupboard, and a big jug from the dishwasher. Then I put the powder in the jug, added some hot water, and poured some into a 2-liter soda bottle we use to feed the babies. I put a rubber tip on the bottle, and walked over to the box where the lamb was bleating hungrily.

I picked him up, and used my thumb and index fingers to pry open his mouth. He caught on right away, and started sucking happily, and was soon done with the whole bottle. Dad walked in right then, carrying dirty towels in his hands. He smiled and looked at the empty bottle. “Someone was hungry, huh?”

"Yeah," I replied, rubbing my eyes sleepily.

"Okay," he said, "why don't you go back to bed? Thanks for your help, kid."

"You're welcome," I said standing up and walking towards my room. I barely had time to kick off my shoes before I collapsed on my bed, exhaustion taking over my body. Knowing how important saving those babies was to the welfare of the ranch was comforting, and with that thought in mind I rolled over and was asleep in a matter of seconds.

About the Author

Emma Morris, age 13

Emma Morris, the award-winning author from Scott Valley Junior High School, found inspiration from her everyday life for her story, *Late Night Lambing*. Having grown up in rural Siskiyou County, and with her experience raising lambs, Emma was excited to get the assignment and thought it was appropriate to write about something she knew well. Emma believes that an intriguing storyline, with accurate facts and good wording, make up a great story. While doing research on her topic Emma not only learned about many aspects of living on a ranch, she also learned that agriculture is a huge source of income for the economy in California and believes everyone should support it. Emma is honored to be recognized as a state winner and hopes her story will encourage other students to learn about and write stories about agriculture.

About the Illustrators

 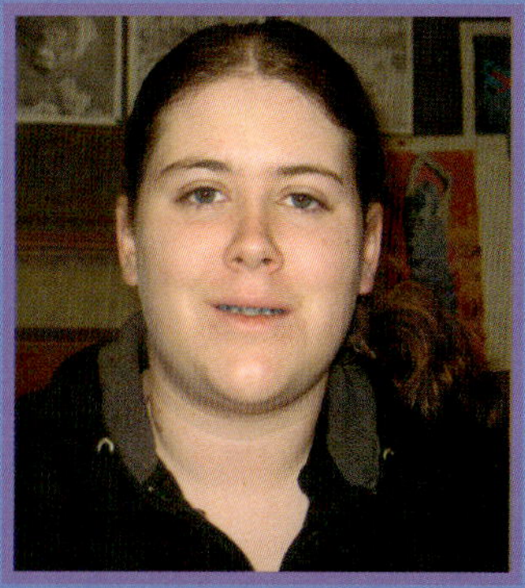

Arden Gabor, Nick Christen and Nicole Jackson
Woodland High School – Woodland Unified School District
Art Instructor: Dawn Abbott

Arden Gabor, Nick Christen and Nicole Jackson are all members of the advanced placement art class at Woodland High School. Woodland High has been providing graphics for the *Imagine this...* Story Writing Contest for three years. This year, students were excited to try a new method for illustrating the story, *Late Night Lambing*. These talented students used a construction paper collage technique to create illustrations with a less obvious realistic effect. They used many contrasting colors and layering techniques to create depth in each scene. They used prisma pencils to add finishing touches. Their goal was to create vibrant pictures that relied on shapes and images to create shadows and depth. This was the first story that Woodland has illustrated with human characters and a more true-to-life storyline. Arden, Nick and Nicole all agreed that it was a challenge to illustrate a story topic they knew virtually nothing about, but gladly accepted the assignment and learned a lot about lambing in the design process.

Illustration by Sheldon High School

IX: The Family Brand

By Braden Whitehouse
7th grade, Grenada Elementary School
Debbi Hoy, Teacher – Siskiyou County

Honorable Mention

I was sitting in class waiting for the final bell of the day. I knew that Grandpa would be waiting out in front of the school with his pickup and stock trailer. Today we were going to ride through his cows out in Quartz Valley. As I rounded the corner of the school I could see the big IX on the front of the stock trailer.

"What does 'IX' stand for?" my buddy Ty asked, looking at the front of Grandpa's trailer.

"It's I-X," I responded, "our family brand!"

After unloading the horses at the Burton Ranch, Grandpa and I set out across the fields to check on his cows. "How many cows should there be?" I asked.

"All together we should have 89 cows and 84 calves," he responded.

"No bulls?" I asked.

"No, they're at the home place, it's calving season right now," Grandpa replied.

After crossing a number of ditches and sumps and opening and closing at least a half dozen gates, we had seen all the cows. Like we always do, I count the calves and Grandpa counts the cows.

"How many did you get?" Grandpa asked as he closed the gate to the last **pasture**.

"Eighty-one," I responded, a little unsure of my count.

"Well good job, because I am missing three cows as well. So there are three pair missing, we will need to find them," Grandpa said.

"How are we going to do that?" I asked.

"We will need to ride through the neighbors' pastures and see if we can find the missing cows."

"Is it all right with the neighbors if we do that?" I asked.

"Sure, it happens all the time, sometimes their cows come into our pasture as well," he answered while we rode.

This didn't seem like it would be very easy I thought to myself. After all, most of Grandpa's cows look the same. They are mostly black, because they are Angus or Angus-cross. Unfortunately, most of the neighbors' cows looked about like Grandpa's I noticed as I looked across the fence while Grandpa got off his horse to get the gate.

I already knew the answer, but I asked anyway. "How are we going to tell our cows from the neighbors'?"

"We will look for those with our brand," Grandpa said.

As we looked through one bunch of cows after another I asked Grandpa why he had the IX brand. He told me he had gotten the brand from his old friend, Emmitt Roberts. Both he and Emmitt liked the simple I and X because it was easy to identify. He explained that every rancher has to register their brand. In addition to the brand design, you have to register a location for the brand on the cow. Grandpa's IX goes on the right hip of all his cows. He says that he likes the hip because it is easy to see from a horse's back.

"So, other people could have the IX brand, but not on the right hip," I said.

"That's right. Emmitt always had the left hip and your aunt has the right rib. Someday your cows will have the IX on the right hip, if you take over my brand." I was so excited just thinking about continuing the family brand.

I noticed as I looked more closely that some cows had more than one brand. Grandpa explained that every time a cow is purchased by someone new they put their brand on the cow. Branding cows has a long history in our country. It is taken very seriously, as it is your only real way to prove ownership of a cow. Grandpa told me that there is a person called a brand inspector that checks the brand of most cows that are sold.

Most ranchers also earmark their cows to help identify them. Grandpa cuts a half moon out of each cow's ear.

After just a few tries I got pretty good at identifying which cows were Grandpa's and not the neighbors'. It didn't take nearly as long as I thought it would to find the three missing pairs and herd them back into the pasture.

As we loaded our horses back into the stock trailer I felt a sense of pride in the family brand and realized for the first time just how important it is to a cattle rancher like Grandpa, and hopefully someday, to me.

Glossary

Bacteria – Living microorganisms. Often used in cheese making to curdle milk and develop aromas and textures.

Coagulate – The act of a liquid transforming into a soft, semi-solid, or solid mass.

Composting – The act of mixing decayed organic matter with soil to improve soil structure and provide nutrients and minerals that help plants grow.

Cultivate – To improve and prepare land for raising crops; to grow or create.

Curds – The part of milk that solidifies when treated with enzymes; used to make cheese.

Drought – A period of dryness that, when prolonged, causes extensive damage to crops or prevents their successful growth.

Ewe – A female sheep.

Fertilizer – Any organic or non-organic substance that is added to soil or water to increase the nutrients available to plants.

Lactic Acid – Produced by beneficial bacteria when metabolizing the lactose in milk for energy during cheese-making.

Lactose – A type of sugar found in milk and milk products.

Lactose Intolerance – The inability to digest milk and some dairy products due to a deficiency of lactase that helps break down milk sugars.

Marmalade – A clear, jelly-like preserve made from the pulp and rind of fruits.

Nutrients – Compounds or elements required by plants for growth and reproduction.

Pasteurizing – Exposing a food to elevated temperatures for a specific amount of time to slow microbial growth.

Pasture – Ground where vegetation grows that is set aside for grazing animals.

Protein – Any of a large group of nitrogenous organic compounds that are essential components of living cells.

Rennet – An enzyme that is used to coagulate one class of protein in milk; used in cheese-making to curdle milk.

Seedling – A young plant that is grown from a seed.

Tryptophan – An essential amino acid that cannot be made by a human body but is found in many proteins such as turkey, eggs and cheese. Often used as a sleep aid because it increases serotonin and/or melatonin levels in the brain.

Vat – A large container used for storing or holding liquids.

Whey – The watery part of milk that separates from the curds in the process of making cheese.

Acknowledgements

California Foundation for
Agriculture in the Classroom

California Foundation for Agriculture in the Classroom (CFAITC) would like to acknowledge the many people who contributed to the success of the 2009 *Imagine this...* Story Writing Contest and the *Imagine this...* book.

Many thanks to...

Imagine this... Regional Coordinators

Sandra Gist-Langiano
Mary Landau
Doni Rosasco
Jacki Zediker

CFAITC Staff

Judy Culbertson, Publisher, *Imagine this...* book
Stephanie Etcheverria, Coordinator and Co-Editor, *Imagine this...* Story Writing Contest and book
Renee Hyatt, Co-Editor, *Imagine this...* book

CFAITC Board of Directors

Kenny Watkins, Chairman
Martha Deichler
Debbie Jacobsen
Jamie Johansson
Bob O'Hollearn
Joe Peters
Jane Roberti

In partnership with
Walmart

About California Foundation for Agriculture in the Classroom

The California Foundation for Agriculture in the Classroom (CFAITC) is a 501(c)(3) nonprofit organization that provides educators with quality free and low-cost materials, training, and information to promote an understanding of California agriculture.

Agriculture in the Classroom is designed to help kindergarten through twelfth grade students acquire the knowledge necessary to become agriculturally literate. Through various programs and resources, educators are encouraged to incorporate agriculture into all subjects and lessons, and to point out the important role it plays in our economy and society.

Why Agriculture?

Our world is a big, beautiful, intricate and often complicated puzzle. In this landscape, every piece plays a critical role, and each relies on another to make the image complete. Agriculture can help give perspective and meaning to the world around us through everyday examples such as plants and animals, demonstrating big-picture issues of water conservation, land use, the environment, and our food supply. Agriculture education creates unique and interactive opportunities to practice important life skills, such as teamwork and communication. Agriculture-based learning can also contribute to a child's increased confidence and improved attitudes and social skills in addition to making a positive impact in their own community.

Imagine this...
Story Writing Contest
Annual Deadline: November 1

Attention Teachers! Use this creative writing contest in your classroom to meet California content standards in reading and writing for grades three through eight.

Students will discover how agriculture affects their everyday lives and their stories will inspire others to appreciate agriculture. State-winning students will have their stories illustrated by high school artists and published in next year's *Imagine this...* book.

Entry Details

- Submit up to five stories from each classroom
- Each story must have an entry form attached
- Stories must be written by one student author; no group entries allowed
- Stories exceeding 750 words will not be selected as state winners
- Stories should not be similar in theme to winning entries from previous years' contests

Winning stories must be

- Titled and original, creative student work
- Fact or fiction
- Appropriate for classroom use
- Grammatically correct
- Related to California agriculture in a positive way
- Typed, preferably, or neatly handwritten
- Written without reference to registered trademarks
- Suitable for illustration

Awards

Please refer to our Web site for a detailed list of the regional- and state-level awards.

www.cfaitc.org

Entry Form

Send story and entry form to your regional coordinator (listed on back).

❑ Male ❑ Female

Student Name ______________________________

Title of Story ________________ Word Count ______

Teacher Name ____________________ Grade ______

(Print first and last name)

Teacher Signature ________________ Class size ______

School Name ______________________________

School Address ______________________________

City, State, Zip ______________________________

School Phone (   ) ______________________________

County ____________ Region Number (see map) ______

Teacher's E-mail ______________________________

Principal ______________________________

(Print first and last name)

Superintendent ______________________________

(Print first and last name)

School District ______________________________

How did you hear about this contest? ________________

Hometown Newspaper ______________________________

❑ Public School ❑ Private School ❑ Home School

Postmark by November 1, annually

Regional Coordinators

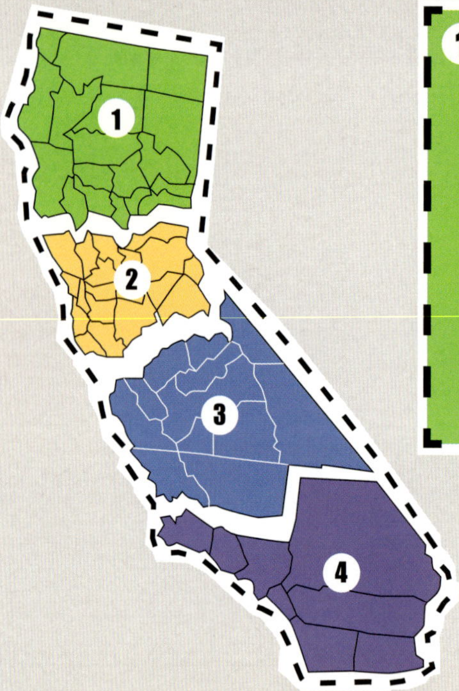

1 **Jacki Zediker**
7132 E. Louie Rd.
Montague, CA 96064
(530) 459-0529

Butte
Colusa
Del Norte
Glenn
Humboldt
Lake
Lassen
Mendocino
Modoc
Nevada
Plumas
Shasta
Sierra
Siskiyou
Sutter
Tehama
Trinity
Yuba

Send story and entry form, postmarked by Nov. 1, to your regional coordinator.

2 **Doni Rosasco**
16002 Hwy. 108, Jamestown, CA 95327
(209) 984-3539

Alameda
Alpine
Amador
Calaveras
Contra Costa
El Dorado
Marin
Napa
Placer
Sacramento
San Francisco
San Joaquin
San Mateo
Santa Clara
Santa Cruz
Solano
Sonoma
Stanislaus
Tuolumne
Yolo

3 **Sandi Gist-Langiano**
P.O. Box 748
Visalia, CA 93279
(559) 732-8301

Fresno
Inyo
Kern
Kings
Madera
Mariposa
Merced
Mono
Monterey
San Benito
San Luis Obispo
Tulare

4 **Mary Landau**
330 East Las Flores Drive
Altadena, CA 91001
(626) 794-4025

Imperial
Los Angeles
Orange
Riverside
San Bernardino
San Diego
Santa Barbara
Ventura